BABIES AT THE ZOO

Hippopotamus Calves

Susan H. Gray

CHERRY LAKE PRESS

Published in the United States of America by
Cherry Lake Publishing
2395 South Huron Parkway, Suite 200, Ann Arbor, MI 48104
www.cherrylakepublishing.com

Content Advisor: Dominique A. Didier, Professor of Biology, Millersville University
Reading Advisor: Marla Conn, MS, Ed, Literacy specialist, Read-Ability, Inc.

Photo credits: ©MANISH DHAPWAL/Shutterstock.com, front cover; ©abxyz/Shutterstock.com, 1, 2; ©ichywong/Shutterstock.com, 4; ©Dario Verdugo/Shutterstock.com, 6; ©Dan Holm/Shutterstock.com, 8; ©Noradoa/Shutterstock.com, 10; ©NiglayNik/Shutterstock.com, 12; ©Patrick Messier/Shutterstock.com, 14; ©Apimook Ph/Shutterstock.com, 16; ©Johan Swanepoel/Shutterstock.com, 18; ©Volodymyr Burdiak/Shutterstock.com, 20

Library of Congress Cataloging-in-Publication Data

Names: Gray, Susan H., author.
Title: Hippopotamus calves / written by Susan H. Gray.
Description: Ann Arbor, Michigan : Cherry Lake Publishing, 2020. | Series: Babies at the zoo | Includes index. | Audience: Grades K-1.
Summary: "Read about hippopotamus calves and how zookeepers take care of them. This level 3 guided reader book includes intriguing facts and adorable photos. Students will develop word recognition and reading skills while learning about how these baby animals learn and grow, what they eat, and how they socialize with each other. Book includes table of contents, glossary, index, author biographies, sidebars, and word list for home and school connection"—Provided by publisher.
Identifiers: LCCN 2019034161 (print) | LCCN 2019034162 (ebook) | ISBN 9781534158931 (hardcover) | ISBN 9781534161238 (paperback) | ISBN 9781534160088 (pdf) | ISBN 9781534162389 (ebook)
Subjects: LCSH: Hippopotamus—Infancy—Juvenile literature. | Zoo animals—Infancy—Juvenile literature.
Classification: LCC QL737.U57 G73257 2020 (print) | LCC QL737.U57 (ebook) | DDC 599.63/513/92—dc23
LC record available at https://lccn.loc.gov/2019034161
LC ebook record available at https://lccn.loc.gov/2019034162

Cherry Lake Publishing would like to acknowledge the work of the Partnership for 21st Century Learning, a Network of Battelle for Kids. Please visit http://www.battelleforkids.org/networks/p21 for more information.

Printed in the United States of America
Corporate Graphics

Table of Contents

About the Author

Susan H. Gray has a master's degree in zoology. She has written more than 150 reference books for children and especially loves writing about animals. Susan lives in Cabot, Arkansas, with her husband, Michael, and many pets.

Name another animal that spends most of its time in water.

Hungry Hippos

Hippopotamuses are big **mammals** that live in Africa. They also live in zoos. These animals spend most of their time in water. Hippos love lakes and rivers.

Hippos leave the water at night. They go looking for food. In Africa, they mostly eat grass. At the zoo, hippos eat hay, **melons**, and vegetables.

A mother hippo is called a cow. Sometimes, a cow gives birth to her baby in the water. She pushes it up to the **surface**. Then it takes its first breath.

A **newborn** can weigh more than 100 pounds (45 kilograms)! The **calf** can walk right away. It yawns when it is sleepy. And sometimes it gets hiccups.

Slime!

Zookeepers take care of hippos at the zoo. They weigh the calves. They clean the hippo swimming pools. They even check the animals' teeth!

What else do you know that is slimy?

Zookeepers know these animals are slimy. The **slime** comes out of the hippo's skin. Slime is a good thing. It keeps hippos from getting sunburns.

Baby Hippo Grows Up

A newborn only drinks milk from its mother. Later, it starts eating grass. By its first birthday, it no longer needs milk.

A hippopotamus stays with its mom for about 8 years. It learns where to find good grass. It learns not to **wander** too far from water.

By this time, the hippo weighs more than a car! It will soon have a baby of its own.

Get ready for a big, hungry baby!

Find Out More

BOOK

Hansen, Grace. *Hippopotamus.* Mankato, MN: Abdo Kids Jumbo, 2017.

WEBSITE

San Diego Zoo Kids—River Hippo
https://kids.sandiegozoo.org/animals/river-hippo
Check out the great photos and quick facts about hippos here.

Glossary

calf (KAF) the young of some animals, including the hippopotamus

mammals (MAM-uhlz) animals that have hair and provide milk to their babies

melons (MEL-uhnz) large fruits with a thick skin or rind

newborn (NOO-born) an animal that was born recently

slime (SLIME) a moist, soft, and slippery thing

surface (SUR-fis) the top layer of something

wander (WAHN-dur) to walk away from where you should be

zookeepers (ZOO-kee-purz) people who take care of zoo animals

Home and School Connection

Use this list of words from the book to help your child become a better reader. Word games and writing activities can help beginning readers reinforce literacy skills.

a
about
Africa
also
and
animal
animals
another
are
at
away
baby
big
birth
birthday
breath
by
calf
called
calves
can
car
care
check
clean
comes
cow
do
drinks
eat
eating
else
even
far
find
first
food
for
from
get
gets
getting
gives
go
good
grass
grows
have
hay
her
hiccups
hippo
hippopotamus
hippopotamuses
hippos
hungry
in
is
it
its
keeps
kilograms
know
lakes
later
learns
leave
live
longer
looking
love
mammals
melons
milk
mom
more
most
mostly
mother
name
needs
newborn
night
no
not
of
only
out
own
pools
pounds
pushes
ready
right
rivers
she
skin
sleepy
slime
slimy
sometimes
soon
spend
spends
starts
stays
sunburns
surface
swimming
take
takes
teeth
than
that
the
their
then
these
they
thing
this
time
to
too
up
vegetables
walk
wander
water
weigh
weighs
what
when
where
will
with
yawns
years
you
zoo
zookeepers
zoos

Fast Facts

Habitat: Lakes, rivers, and swamps

Range: Africa

Average Length: 9.5 to 14 feet (3 to 4 meters)

Average Weight: 3,000 to 8,000 pounds (1,361 to 3,629 kilograms)

Life Span: About 40 to 45 years

Anatomy: Eyes, ears, and nostrils are on the top of the head. A hippo that is mostly underwater can see, hear, and breathe.

Behavior: Hippos often make loud sounds underwater. These sounds can warn other hippos of danger.

Index